Creating a Culture of Integrity

Business Ethics for the 21st Century

One Stone Advisors
www.onestoneadvisors.com

Andrea Spencer-Cooke
Partner One Stone Advisors
andrea@onestoneadvisors.com

Fran van Dijk
Partner One Stone Advisors
fran@onestoneadvisors.com

Published by Greenleaf Publishing Limited
Aizlewood's Mill
Nursery Street
Sheffield S3 8G
UK
www.greenleaf-publishing.com

Printed and bound by Printondemand-worldwide.com, UK

ISBN 978-1-910174-60-9 (eBook-ePub)
ISBN 978-1-910174-61-6 (eBook-PDF)
ISBN 978-1-910174-59-3 (Paperback)

A catalogue record for this title is available from the British Library.

Page design and typesetting by Alison Rayner
Cover by Orest Viruta

DōShorts

Dō Sustainability is the publisher of **DōShorts:** short, high-value business guides that distill sustainability best practice and business insights for busy, results-driven professionals. Each DōShort can be read in 90 minutes.

New and forthcoming DōShorts – stay up to date

We publish new DōShorts each month. The best way to keep up to date? Sign up to our short, monthly newsletter. Go to **www.dosustainability.com/newsletter** to sign up to the Dō Newsletter. Some of our latest and forthcoming titles include:

- ***Lobbying for Good: How Business Advocacy Can Accelerate the Delivery of a Sustainable Economy*** Paul Monaghan & Philip Monaghan
- ***Creating Employee Champions: How to Drive Business Success Through Sustainability Engagement Training*** Joanna M. Sullivan
- ***Smart Engagement: Why, What, Who and How*** John Aston & Alan Knight
- ***How to Produce a Sustainability Report*** Kye Gbangbola & Nicole Lawler
- ***Strategic Sustainable Procurement: An Overview of Law and Best Practice for the Public and Private Sectors*** Colleen Theron & Malcolm Dowden
- ***The Reputation Risk Handbook: Surviving and Thriving in the Age of Hyper-Transparency*** Andrea Bonime-Blanc
- ***Business Strategy for Water Challenges: From Risk to Opportunity*** Stuart Orr and Guy Pegram

- ***Accelerating Sustainability Using the 80/20 Rule*** Gareth Kane
- ***The Guide to the Circular Economy: Capturing Value and Managing Material Risk*** Dustin Benton, Jonny Hazell and Julie Hill
- ***PR 2.0: How Digital Media Can Help You Build a Sustainable Brand*** John Friedman
- ***Valuing Natural Capital: Futureproofing Business and Finance*** Dorothy Maxwell
- ***Storytelling for Sustainability: Deepening the Case for Change*** Jeff Leinaweaver
- ***Beyond Certification*** Scott Poynton
- ***21st Century Growth: Beyond the Water-Energy-Food Nexus*** Will Sarni

Subscriptions

In addition to individual sales of our ebooks, we now offer subscriptions. Access 60+ ebooks for the price of 6 with a personal subscription to our full e-library. Institutional subscriptions are also available for your staff or students. Visit **www.dosustainability.com/books/subscriptions** or email **veruschka@dosustainability.com**

Write for us, or suggest a DōShort

Please visit **www.dosustainability.com** for our full publishing programme. If you don't find what you need, write for us! Or suggest a DōShort on our website. We look forward to hearing from you.

Abstract

FOR COMPANIES, UNETHICAL BUSINESS PRACTICES such as bribery and corruption pose major business risks. Such conduct can result in significant financial costs in the form of fines, reputational damage, lost business opportunity and – increasingly – criminal or civil charges. Ethical conduct is therefore a critical governance and reputational concern and organisations have responded by putting in place rigorous formal integrity and compliance programs to set out and enforce clear standards for ethical business practice. But to really embed good conduct requires more than compliance alone. Companies also need to create an enduring culture of integrity that establishes doing the right thing as the cultural norm across the organisation. This book identifies the key actions needed to foster this cultural shift. As well as best-practice corporate examples of embedding ethical behaviour, it provides a toolkit that sustainability managers and compliance officers can draw on to engage everyone in the company around this critical issue.

About the Authors

ANDREA SPENCER-COOKE has been involved in shaping sustainable business thinking and practice at leading companies for over two decades, including as editor of *Tomorrow – Global Sustainable Business* magazine and programme head at SustainAbility Ltd. Her focus is on corporate leadership and accountability, with a particular interest in changing management mindsets around business values and role in society. Andrea is a trained Primary Ethics teacher, with a background in social anthropology and responsible business.

FRAN VAN DIJK is a sustainability strategist working with leading companies to demonstrate the benefits of sustainability and turn risk into opportunity. Formerly a senior consultant and Head, Sustainable Lifestyles with SustainAbility Ltd, her clients have included BT, Procter & Gamble, RB, Gambro, Electrolux and Tetra Pak. In 2010, she was part of the ENDS Carbon team behind the FTSE CDP Carbon Strategy Index Series. Fran served on the board of the Macaulay Land Research Institute for seven years.

Acknowledgments

MANY PEOPLE HAVE HELPED TO SHAPE THIS BOOK through a shared interest in finding better ways to do business and we'd like to acknowledge the inspiration we've drawn from them (see Chapter 10). In particular, we would like to thank those we consulted and who generously made time to share their expertise and insights in business ethics: Tim Erblich, Ethisphere; Simon Webley, Institute of Business Ethics; Kirk Hanson, Markkula Center for Applied Ethics, University of Santa Clara; John Haidt, NYU-Stern School of Business; John Neil and Simon Longstaff, St James Ethics Centre; Jacqueline Mees, University of Sydney and Bill O'Rourke, Wheatley Institution, Brigham Young University. A very special thanks goes to Alex Slippen and the Ethics Resource Center (ERC), for allowing us to use and reproduce ERC research, and to Kirk Hanson for sharing his first-rate MOOC on Creating an Ethical Corporate Culture.

We're also extremely grateful to the companies who talked with us so openly about their experience in embedding an ethical culture: Audrey Tillman, Aflac; Sam Al Jayousi, Carillion; Malin Ekefalk, Electrolux; Sefton Laing, RBS; Andreas Follér, Scania; and Michaela Ahlberg, TeliaSonera. Both Jens Schlyter and Helena Hagberg, Swedish Ministry of Enterprise, contributed a valuable public sector perspective. A number of One Stone client companies feature in the book and we'd like to underline again what a pleasure it is to work with you.

Finally, we'd like to recognise the talented Amy Brown for her help with research, interviews, editing and her enthusiasm and support; and Astrid von Schmeling at Purple Ivy, whose excellent interviews from Sweden enrich the book – *tack så mycket*! Comments on early drafts by Sarah Holloway and April Streeter were extremely helpful – any errors or omissions are ours alone. Finally, a big thank you to Nick Bellorini and the DōShorts team for making the book happen, and to our wonderful families for putting up with the long hours.

This book is dedicated to all those in business who do the right thing – you know who you are.

Contents

CHAPTER 1

Introduction

ETHICS IS A GROWING BUSINESS ISSUE. You can barely open a newspaper today or turn on the TV without hearing something about yet another company paying the price for poor employee judgment, bad conduct or inadequate oversight. In a 24-hour news cycle world, nothing stays hidden for long – quite literally in the case of Petrobras, Brazil's state-owned oil company, whose multi-billion dollar corruption probe has even led to calls for a suspect to be exhumed.[1]

The penalty for such misconduct is high. Back in 2008, German engineering giant Siemens made corporate history when it paid US$1.6 billion to settle charges of bribery across its global operations.[2] That has since been dwarfed by banking fines in the US and UK: $1.9 billion to HSBC, $2.6 billion to Credit Suisse, and a monumental $8.9 billion to BNP Paribas. Their transgressions? Everything from mis-selling insurance and helping customers dodge tax to rigging forex markets and violating sanctions.[3]

The actual cost of irresponsible decision-making is even higher. UBS losses in the sub-prime mortgage crisis totalled US$17 billion – the largest in Swiss corporate history – and could have been prevented with a different risk culture.[4] Studies put the cost of reputational damage from ethics scandals at up to seven times the original fine in market value.[5] World soccer federation FIFA is learning the hard way what graft does to your reputation, but it could be some time before the final price-tag for its appalling governance is known.

A shared burden

Added up, the burden of unethical business conduct on the economy as a whole is mindboggling. The B20 Anti-Corruption Working Group estimates that corruption consumes 3% of global GDP each year, and costs more than US$2.6 trillion.[6] It reduces government revenue, undermines growth and development, and increases the costs of doing business by up to 10% on average according to the World Economic Forum and OECD.[7] The rationale for fighting corruption isn't just a question of ethics – we simply can't afford it.

The cost isn't only economic. Corruption undermines the rule of law and the power of democracy. As Harriet Kemp, Head of Development at the Institute of Business Ethics (IBE) puts it, 'corruption is not a victimless crime; it leads to decisions being made for the wrong reasons. Contracts are awarded because of kickbacks and not whether they're the best value for the community. Corruption costs people freedom, health and human rights and, in the worst cases, their lives.'[8]

Tightening the net

Little wonder that business 'conduct risk' is a red-hot topic for regulators. Around the world, standards and regulations are being developed and revised to combat unethical business practice more effectively, from the US Foreign Corrupt Practices Act (FCPA) and UK Bribery Act to China's Article 164. The OECD Anti-Bribery Convention, legally binding in over 40 countries, also makes it a crime to bribe foreign public officials in international business transactions.

As well as legal frameworks, voluntary anti-corruption initiatives are proliferating. The United Nations Global Compact (UNGC), whose tenth principle targets ethical business conduct, now counts among its signatories over 8000 businesses across 145 countries. At industry level, too, bodies like the UK Banking Standards Review Council (BSRC) are being created to improve behaviour and restore public trust, while the Governor of the Bank of England Mark Carney has called for global standards to tackle the sector's 'culture of impunity'.[9]

The upshot at individual company level is a race to the top to foster good conduct by embedding effective ethics and compliance (E&C) frameworks that prevent, detect and respond to ethical violations. But no E&C programme alone can guarantee that everyone will always do the right thing. Unless it's accompanied by values-driven behaviour change that is aligned, integrated and reinforced at individual, organisational and systems levels, your E&C initiatives will struggle to make a lasting impact (see Chapter 9).

Rebuilding trust

At stake here for business is not just a hit to the bottom line in the form of fines, but a much wider issue of trust, credibility and organisational well-being. Most companies, most employees want to do the right thing. But in the words of Sam Al Jayousi, Group Compliance Manager at Carillion, one of the UK's leading integrated support services companies, 'reputational damage is contaminating'.[10] A few rogue players in a company or industry can be enough to tarnish everybody – can you name a 'clean' bank?

When public trust takes a hit, a company's licence to operate is jeopardised. Regulators step in, employee morale tumbles, brand reputation suffers, customer loyalty weakens, investors and business partners flee and it becomes harder - much harder - to attract and retain top talent.

Conversely when a corporation gets it right, by promoting, supporting and celebrating personal and organisational integrity and empowering employees to 'do the right thing', not only is conduct risk reduced, but the foundations are laid for long-term business success. Reshaping your business culture is the best way of preventing ethical dilemmas from becoming ethics lapses in the first place.

> ***"In the wake of the corruption scandal, Siemens was gripped by a mood of "something needs to happen." Staff were shocked, but that gave rise to a sense of purpose and new beginning. We need to feel that again - albeit under more auspicious circumstances. Ultimately, strategy papers don't make or break the future and sustained success of a company. Its corporate culture does."***
> **JOE KAESER, CEO SIEMENS**[11]

Why read this book?

Drawing on examples from pioneering companies around the world, this book demonstrates the importance of fostering a culture of integrity as a foundation for ongoing corporate anti-corruption initiatives.

Without a strong ethical culture companies have to rely on the compliance system to do all the heavy lifting. Investing in integrity means shifting from a 'don't get caught' to a 'right thing to do' mindset that can positively reinforce compliance, and boost trust and openness.

Step one is adopting standards and setting up a compliance system to enforce them. Step two is about creating an enduring culture of integrity that embeds good business conduct as the cultural norm across the organisation. This book shows you how.

As well as examining the role of policies and codes (Chapter 3), it explores how recruitment, training, management processes and tone from the top can be harnessed to incentivise ethical practices and build a strong organisational culture of trust, openness and integrity (Chapters 4–7). It gives tips on what to do when things go wrong (Chapter 8). And it describes how, in a culturally diverse world, companies and stakeholders can work together to change the unwritten rules that work against us doing the right thing (Chapter 9).

CHAPTER 2

Moving Beyond Compliance

IF YOUR COMPANY HAS ALREADY SET UP an Ethics and Compliance (E&C) system, you've taken the first important step towards embedding ethics.

Basic E&C checklist

Although the E&C function is relatively new, there is broad consensus on the elements of an effective programme (see Box 1, overleaf). Front and centre are risk assessment, clear codes and policies, and putting in place the right organisational roles, structures and resources. Next, come training and communication – vital for on-boarding employees – proper internal controls and a safe, anonymous, third-party hotline for reporting violations. Finally, penalties and regular audit and reporting cycles help to ensure that standards are upheld.

In addition, the US Department of Justice and US Securities and Exchange Commission list four key things they see as hallmarks of an effective E&C programme:[13]

- Commitment from senior management.
- Incentives (i.e. personal evaluations and promotions, rewards for ethics and compliance leadership) and disciplinary measures.

- Due diligence around payments to third parties, such as agents, consultants and distributors.
- Merger and acquisition (M&A) due diligence and integration.

Properly implemented, an E&C programme that ticks all these boxes will go a long way towards embedding ethical conduct. But without a culture that supports it, even the best E&C system will struggle to deliver.

BOX 1. Elements of an effective E&C programme[12]

1. Periodic and targeted E&C risk assessment.
2. Code of Conduct (CoC) and related system or framework of policies.
3. Chief ethics and/or compliance officer (CECO) with sufficient resources and budget.
4. CECO access and reporting to the management and board.
5. Appropriate ethics and compliance training and communications.
6. System of internal controls and proper delegation of approval authority.
7. Helpline/hotline system with anonymous reporting options.
8. A consistent system of internal discipline.
9. Periodic auditing, monitoring and evaluating of E&C programme.

Prevention is better than cure

A company's conduct risk is determined by the behaviour of its employees. To sustain high standards of ethical conduct, you need everybody in your organisation to buy into that goal. Preventing corruption, bribery or fraud *before* it happens is far better than catching people after the fact. By the time the ref blows the whistle, it's too late – as FIFA is learning.

Rigorous E&C programmes can certainly drive people to act ethically, but it may just be in order to follow the rules, or out of fear of getting caught, rather than any genuine desire to do the right thing. Why isn't that enough? First, because policies and regulations tend to draw on past experience to prescribe *minimum* expected standards so are always playing catch-up. Second, they tend to focus on common scenarios, yet in a rapidly changing global business landscape they can't cover every eventuality, predict every situation or cater to the true complexity of ethical 'grey areas'.

Integrity doesn't just stem from following the letter of the law: it comes from empowering employees to consistently reflect on and choose to uphold ethical principles. To do the right thing every time in the face of 'frontline' business pressures like winning a big contract in a new market, your people need to have the ethical sensitivity to make the right decisions, not just follow a policy script. As corporate culture change consultant and speaker Micah Solomon puts it: 'If employees are only doing things right because you spelled every little thing out, even if you do so very, very elegantly, you haven't created a culture, and you haven't created an approach that is sustainable. A culture is a living thing, powered by and kept up to date by the people who are encouraged to be, in a meaningful way, part of it.'[14]

Why culture matters

Company culture and norms are critical in explaining why and how unethical behaviour occurs. Corruption takes root where opportunity and culture let it: industries become prone to practices like bribery because collective pressures end up shaping how business gets done, 'normalising' unethical behaviour. OECD analysis of foreign bribery enforcement actions, for example, has found that four sectors account for 75% of all cases: extractive, construction, transportation and storage, and information and communication.[15]

When a sector's culture drifts grossly out of alignment with stakeholder values and expectations – as in banking, for example – ethical problems arise and profound change is required. Federal Reserve Bank of New York President William Dudley has argued that ethics lapses by so-called rogue bankers like Nick Leeson or the 'London Whale' shouldn't be seen 'as isolated actions by deviant employees, but as evidence of failure by senior managers – from the boardroom to the executive suite – to orient bank culture properly'. His answer? To 'infuse their organisations with an ethos of compliance and integrity'.[16] Invest in changing the culture, in other words, and good conduct will follow.

'You can sort out systems to a point and introduce policies and processes to prevent ethical conduct problems from happening, but unless you have people onboard from a business ethics point of view, bad things can still happen,' Sefton Laing, Head of Sustainability Services at RBS confirms. 'You have to address the conduct of your people and the culture of ethics that means that all the thousands of small decisions taken every day are the right decisions. You can have great programmes, teams, reporting, good governance, but it's the small decisions and conduct that count.'[17]

Long-running research by the US-based Ethics Resource Center (ERC) supports the central role of culture in predicting ethical conduct. Its 2013 national survey of the US workforce found that frequency of misconduct mirrors the strength of a company's ethics culture. In what ERC defines as 'weak' ethical cultures, misconduct was four times more evident (see Figure 1) with more than four out of five (82%) misdeeds happening repeatedly. In companies with 'strong' ethics cultures, misconduct was

FIGURE 1. Misconduct declines as ethics culture improves.

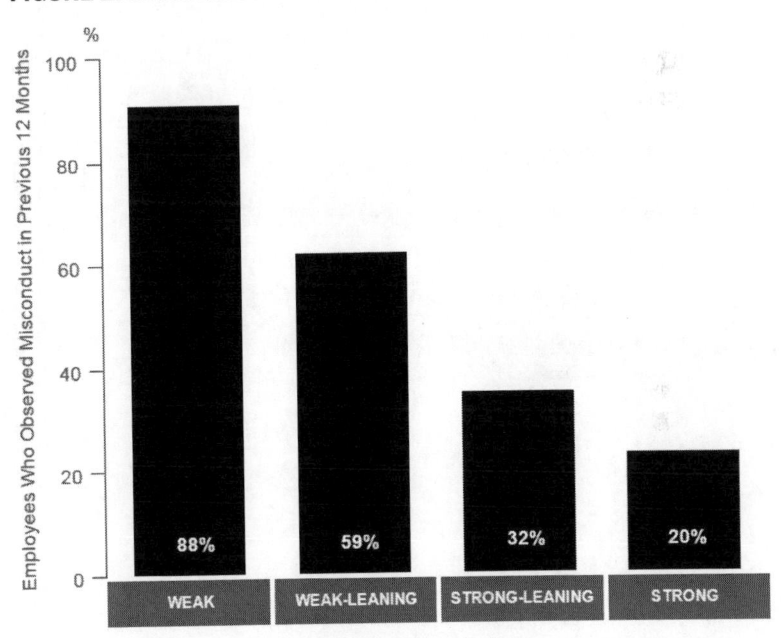

SOURCE: Ethics Resource Center, 2014. National Business Ethics Survey of the US Workforce (NBES 2013), Arlington: ERC.

not only less frequent, but in 60% of cases it was a one-off incident rather than standard operating practice.[18]

The key to a strong ethical culture is clear corporate values. Values shape how things get done, and how you do things in turn creates your culture. Empower your employees with a strong, shared foundation of ethical values and support them to live these every day and your culture will be working with your E&C programme, not against it.

Reasons to move beyond compliance

- E&C programmes alone won't deliver a strong ethical culture. Your people need ethical sensitivity to make the right decisions every day.
- Policies and regulations that set out *minimum* expected standards can't capture the full complexity of ethical 'grey' areas and are always playing catch-up. Strong values can help fill the gap.

CHAPTER 3

Culture and Values

SO YOU HAVE AN E&C PROGRAMME IN PLACE. Check. You 'get' that culture is an important driver of ethical conduct. Check. But what exactly *is* culture and how do you go about ensuring that yours can deliver?

What is corporate culture?

Culture is a pretty vague term. Put simply, it's 'the way we do things here', the dominant ethos. It is the shared habits, accepted attitudes and collective actions that result from the values you and your colleagues hold. And it's a vital ingredient for corporate success.

'Culture is something intangible', says Andreas Follér, Sustainability Manager at heavy vehicles manufacturer Scania. 'It's what we do and say, how we behave towards each other.' Scania prides itself on having a strong culture of responsibility towards others – of doing things right. 'It helps minimise wrongdoing', Follér stresses, 'but more importantly I think a company with a strong culture can be more efficient and agile. It gives us a common foundation of understanding. A company with a strong culture builds trust with others, which in turn leads to better results.'[19]

Culture emanates from the clusters of behaviour and beliefs of the individuals who make up an organisation. It's shaped by the systems and policies you put in place – what you aim for, communicate, model, support and reward. At its root, it boils down to two things: 1) the organisation's purpose – why you're in business, and 2) your values – what you choose to prioritise. These are the guiding forces that determine how your people act and treat others every day.

Strong core values

At the outset, culture is a product of a company's founding values, its so-called heritage. Ethical cultures are ones where principles like trust, fairness, integrity and respect are baked into company mission. Good examples were former British Quaker firms like Cadbury's and Rowntree's, set up to be ethical as well as profitable and whose family-ownership meant they could embody personal values.[20]

Defining values and creating a culture that guides employees' actions, attitudes and behaviour takes time. Ideally, it becomes a shared journey of co-creation that everybody buys into. It isn't enough to communicate top-down slogans and procedures: employees have to align themselves with – and be motivated to live by – your organisational values.

According to Simon Webley, Research Director at the Institute of Business Ethics (IBE), we articulate our values in terms of how we see our roles and corresponding obligations to others – how we treat stakeholders and the environment, in other words. 'Any commercial organisation has at least five overarching stakeholder groups', he argues. 'These five "umbrella" groups are staff, customers/clients, suppliers/service providers, shareholders and/or other providers of finance, and the broader societal communities in which an organisation operates or on which it impacts.'[21]

A statement of core values should set out who your stakeholders are and your aspired behaviour towards them. Identifying specifically how you want your values to apply makes it easier to cascade them through the organisation and translate what they mean for each role and function. That in turn makes it easier to identify what ethical dilemmas or values conflicts could arise in each role.

'The Johnson & Johnson Credo is the granddaddy of values statements', says Kirk Hanson, Executive Director of the Markkula Center for Applied Ethics at Santa Clara University.[22] It spells out the obligations the company has to its stakeholders and the communities where it operates (see Box 2).

BOX 2: Johnson & Johnson Credo[23]

'We live by a set of shared values set out in Our Credo. This guides our actions and behaviours and helps us to create an environment we're proud of. We believe that employees who are valued, respected and treated fairly bring passion and focus to their work and make important contributions. And we also believe that by embracing the ethical principles and responsibilities in Our Credo and the Health Care Compliance code, we will make sound decisions that benefit everyone.'

'Robert Wood Johnson, former chairman from 1932 to 1963 and a member of the Company's founding family, crafted Our Credo himself in 1943, just before Johnson & Johnson became a publicly traded company. This was long before anyone ever heard the term "corporate social responsibility". Our Credo is more than just a moral compass. We believe it's a recipe for business success. The fact that Johnson & Johnson is one of only a handful of companies that have flourished through more than a century of change is proof of that.'

Living your values

To remain strong, core values have to be lived. 'The actions of a business *are* its value statement', says Joel Peterson, chairman of JetBlue Airways. 'Actions speak volumes about what really matters most – not what should matter, not what we wish mattered, but what really does matter to us. They swamp mission statements, speeches or memos, and they eclipse intentions.'[24]

If they aren't actively maintained, over time values can change, become distorted or diluted, depending on the characteristics of the leadership team and people who work there. Actions are also influenced by the wider business ecosystem a company operates in – the standards and expectations at play across its value chain and industry sector.

These 'unwritten' rules mean that the actual culture in an organisation may differ markedly from its stated values. Over time, too, unethical behaviour can become so engrained in company culture that it is not considered unethical at all.[25]

A 2015 survey of 1200 finance professionals, for example, found widespread evidence of an industry culture of secrecy and misconduct: almost half thought competitors had behaved unethically or illegally to gain a market edge, nearly a quarter thought their colleagues had done so, and a fifth had first-hand knowledge of wrongdoing in their workplace.[26]

'Our behaviour is influenced by the norms that we believe exist in the industry, the norms that we believe exist in the organisation', explains study co-author Ann Tenbrunsel, David E. Gallo Professor of Business Ethics at the Mendoza College of Business at the University of Notre Dame. '[If] everybody else is doing this, we also know from psychological research on peer pressure that I will be more likely to do it myself.'[27]

Managing your values

The way to create and sustain a strong ethical culture is to proactively manage your values. To ensure you don't become the next Enron, ERC has some tips for turning strong values into strong culture:[28]

1. Conduct a corporate culture assessment of prevailing values, attitudes, perceptions, standards of conduct, pressures to commit misconduct, communications, vulnerabilities and risks.
2. Pay special attention to how well the board, senior leadership, employees and key stakeholders have internalised and act out your corporate values.
3. Choose to live your corporate values by publicly committing to being an ethical organisation – don't just print, post and pray.
4. Create a robust ethics infrastructure that bakes ethical conduct into corporate systems via key performance indicators (KPIs), incentives and targets.
5. Talk about values often and back it up with open lines of communication.

US insurance provider Aflac is the only company in its sector to have made it onto Ethisphere's list of World's Most Ethical Companies for nine years straight. One of the reasons for its unique success has been building a 'spirit of trust' over time. Culture has been key (see Box 3).

BOX 3: Act like a duck[29]

The 2013 Corporate Citizenship Report of US insurer Aflac is entitled 'What It Means to Act Like a Duck.' The title refers to the iconic white duck that is the company's marketing face and voice. 'To the uninitiated that seems silly', writes Chairman and CEO Dan Amos, 'but to those familiar with Aflac, "acting like a duck" is another way of capturing our corporate culture, focusing attention on doing things the Aflac way.'

The Aflac Way is a hardcover book given to every new employee. It contains seven commitments founded on the company's business approach – such as 'communicate regularly', 'treat everyone with respect and care' and 'shoot straight'. Being there for the customer in their time of need sits at the heart.

Aflac has gone the extra step of aligning the company's brand promise with employee conduct. When you act, you *are* the company – a powerful way to bring home the importance of personal integrity in creating corporate value.

'It would be easy to say that we've employed a multitude of best practices that have led to an honest and ethical culture at Aflac', says Audrey Tillman, Executive Vice-President and General Counsel. 'We do have those practices in place, particularly when it comes to our compliance functions. But the other reality is that at Aflac, integrity has always been instilled in our culture. If you walk around our buildings you'll see reminders of that on our walls, in our conference rooms. Unethical behaviour is not tolerated at Aflac.'[30]

When it comes to responsible conduct, it's this commitment to shaping organisational values and culture that sorts the leaders from the rest.

Five tips to strengthen your culture

- Assess the state of your culture and use the findings to reinforce a clear purpose and shared set of values for your organisation.
- Centre your values on how you aspire to treat your five key stakeholder groups.
- Go public: keep values strong by letting people know what you stand for.
- Keep daily actions aligned with your values by putting in place systems to actively manage them.
- Connect the dots for employees between their personal integrity and your brand value to show how important they are to your culture.

CHAPTER 4

Setting Up Support Systems

ONCE YOU'RE CLEAR ON WHAT YOUR COMPANY STANDS FOR, it's time to set up the necessary support systems to ensure the people in your organisation actually live these values. Applying values can be hard – especially in the face of intense pressure to make money or seal a deal – so you need to make sure the signals and support you give employees make it as easy as possible for them to make the right choices.

Systems matter. The processes you put in place are the key mechanism for translating your values into organisational practices. Making sure they reinforce your ethical standards is vital. This means harnessing the power of every department in the organisation to push the message out, embed it, measure its impact and improve it. The first step is making sure you give your E&C function the right reach.

Supercharging the E&C function

Having a dedicated function to manage how a company does business and conducts its relationships is still relatively new, underlines Timothy Erblich, CEO of US-based Ethisphere, and its remit, resourcing and position in the chain of command varies widely. 'Now companies often have a position like Chief Ethics Officer – a few years ago that didn't exist. Today they often report directly to the CEO with a dotted line to the board. In the past they would typically have reported to the Legal function.'[31]

Having the ear of the CEO or board ensures that ethics is taken seriously. How you define the purpose of E&C in the organisation also matters. An IBE research report on the role of E&C practitioners in promoting ethical conduct has identified three primary 'domains of activity':[32]

- *Custodianship* – a compliance or policing focus, with an emphasis on enforcement of standards.
- *Advocacy* – a challenger role, raising difficult issues and promoting dialogue.
- *Innovation* – a change management role, working as a partner in transforming business processes and risk profiles.

Enforcement is important, but it's when you combine it with advocacy and innovation that you get real traction and organisational transformation. Challengers pose the tricky questions and point out red flags before issues arise. Innovators work to reshape the business in line with purpose and values to design out ethical hazard. 'It's our job to be irritating, to be the gravel in the shoe', laughs Michaela Ahlberg, Chief Ethics and Compliance Officer at Swedish telecom giant TeliaSonera, 'but I also try to be pleasant.'[33]

The trick is to ensure that your E&C support system covers all three aspects – compliance, advocacy and innovation – by working alongside other functions. Carillion is a company that seems to be getting this right (see Box 4).

BOX 4: The compliance manager as enabler[34]

Carillion Group Compliance Manager Sam Al Jayousi has top access and a role that straddles advocacy and innovation. 'I get to see the board whenever I want. I have the liberty to say when the management team needs to take action, address something. Tone from the top is not just when the CEO says the right thing, but when the compliance function has the ability to access senior people – it's about openness and engagement – having the right corporate governance structure in place.'

Al Jayousi doesn't see his job as ensuring compliance across the company. 'Compliance and ethics is the responsibility of every person in the company. My job is to raise awareness, monitor and report back, and co-ordinate the compliance activities. It's to ensure the managing director of every business is aware of the policies and procedures.

He reports to the board on what's happening in the company, where the pinch-points are and where to focus more attention. 'I go to the General Counsel, the CEO, the finance director and tell them about risks and put programmes in place to mitigate these. I get involved in the whistle-blowing committee, in any new merger or acquisition, with board meetings where needed and see all the board papers – I'm a big part of the corporate governance system of the company.'

Breaking down silos

If ethics is treated as an add-on, it will be an afterthought. To become engrained in the culture, a sense of ownership must spread across business functions and activities. That calls for a cross-functional, team-based approach. A vital task of the ethics manager is therefore to get every level of the organisation to identify how values apply. You can then get to work on embedding expectations by educating employees, getting leaders to model the right behaviours, and tying performance rewards to standards and values.

> ***“You can't hang culture change on any one department, it has to be across the board.”*** **SEFTON LAING, HEAD OF SUSTAINABILITY SERVICES, RBS**[35]

A good approach is to put in place a steering group of senior people from every discipline who meet regularly to review progress, discuss approach, policies and procedures, then go back and push this down through the business. You don't want it to become a second job: to make it as easy as possible, try to piggyback on existing initiatives, like quality or health and safety (H&S).

That rationale is sound, agrees Carillion's Al Jayousi: ‘H&S is always top of the agenda for us, so we put E&C on the agenda, too. If you as manager are already giving an H&S talk for the guys on site, speak about the "method statement" – but also every couple of weeks talk about the whistle-blowing line or gifts and hospitality policy.’[36]

By making ethics a cross-functional issue you create synergy, tapping into diverse expertise – from human resources and legal to internal audit and sustainability – and building engagement throughout the company. This integrated governance model has helped Electrolux drive ethics throughout the home appliance Group. 'It works well', states Malin Ekefalk, Director of Corporate Social Responsibility (CSR). 'All functions are represented at steering group and working group level. We developed the programme together. Ethics needs to be part of the leadership model and how you recruit and manage employees in general.'[37]

Appoint ambassadors

A number of companies have also set up networks of 'ethics ambassadors' to create a multiplier effect and cascade the message through the organisation. Former President of Alcoa Russia Bill O'Rourke, now an Ethics Fellow at the Wheatley Institution at Brigham Young University, describes how the company created an Integrity Champion Network. 'We had forty individuals around the world who were long-term employees, usually in a leadership position, who had the utmost integrity and openness – people you could trust. They'd get on the phone once a month and be available to anyone in the organisation to talk about ethical issues.'[38]

Appointing some handpicked champions with a direct link to the ethics function at HQ is very good practice. They can serve as your eyes and ears to make sure training is done and lessons shared. It's a model that works to break down silos and embed ethics at different levels of the organisation while delivering three key benefits:

- The compliance function doesn't need to do all the heavy lifting – having all units pulling in the same direction creates synergies and virtuous circles.

- Being able to plug into the expertise of multiple functions and piggy-back on existing systems gets the message out more effectively.
- Creating a network of cross-functional ethics ambassadors or integrity champions accelerates the change process.

Communicating across cultures

One of the most commonly voiced challenges of rolling out an effective E&C programme and getting employees on board is dealing with different cultures. Barriers include language, attitudes and customs as well as diverse regulatory frameworks; but don't use cultural diversity as an excuse. The same standards should apply to everyone.

> ***"You need to recognise that cultures are different, but when it comes to doing the right thing, I think they're aligned. Don't allow culture to make you behave differently."*** **BILL O'ROURKE, FORMER PRESIDENT ALCOA RUSSIA**[39]

A common example of a difference in culture is willingness to use the whistleblower hotline. Under-reporting – where people don't trust authority, don't want to lose face or are afraid – is a major concern. Yet when you're dealing with national cultures of reluctance to report it's hard to drive change.

'Companies struggle with the hundreds of years of regional and cultural differences embedded in places like Asia, in Latin America, in Africa, where people don't want to raise their hands', sympathises Ethisphere CEO Timothy Erblich. 'The best examples of ethical companies are those that are in 60–70 countries but have one set of rules, one set of ways they operate.'[40]

General Electric (GE) deals with this by offering five different pathways for employees to flag ethical concerns: via the compliance, HR or legal departments, through a manager, or using variety of channels to anonymously contact an ombudsman. Encouragingly, the number of calls at GE has risen, a sign the approach works.[41]

This is an evolving area (see Chapter 9), but emerging good practice points to having one set of standards, keeping the message simple and applied, and working with local staff to get the message across in a variety of ways that mesh with different world views (see Box 5).

BOX 5: Working across cultures[42]

With around 60,000 employees and over 60% of its manufacturing operations in emerging markets, appliance manufacturer Electrolux understands the challenges of rolling out a global ethics programme. CSR Director Malin Ekefalk shares some insights:

'Use group policy to set out a framework of minimum standards. Then you inform, educate and nurture through dialogue. Make it relevant, minimise the policy talk, focus on "what does this mean to me" via examples and keep at it continuously. Have information available in local languages to inform people. Give them examples of how cases are dealt with and the outcomes, including where there's been too little information to investigate.'

A key learning is to keep it simple. 'We had many different components in the training programme. This meant we could pick and select for each market – which was good – but it made translation and development really difficult as booklets, slide-sets, e-learning and the portal all had to be translated into local languages. So keep it simple and pick out the key things, especially if working with limited resources.'

By putting in place strong support systems, you give your E&C programme the best chance of success. Then you can shift attention onto how to model, educate for and incentivise desired behaviours and attitudes, wherever you do business.

Five tips on support systems

- Give your E&C function access to the top – having the board's ear drives ethics up the organisational agenda.
- Make sure you consider all three aspects of the E&C function: compliance, advocacy and innovation.
- Engage a cross-functional team to maximise synergies, with ambassadors to help push the message down.
- Have one set of rules worldwide, but adjust communication to accommodate cultural differences.
- Review existing systems to spot how you can improve the way your organisation models, educates about and rewards right behaviours.

CHAPTER 5

Model: Leading by Example

IT'S A CLICHÉ, BUT WHEN IT COMES TO ETHICAL CULTURE, tone from the top – or how the most senior people in your organisation act – really does count. Leaders set the example. They determine direction, goals and priorities. They make important decisions and choose who and what to reward. And when things go wrong, they determine the consequences. Getting the role models and authority figures in your company to walk the talk may be the single most important thing you can do to build your culture of integrity.

How not to lead

Figures show there's work to do to get corporate leaders to live by the high ethical standards expected of them. As people rise up the hierarchy the stakes get bigger, and so do the pressures and temptations. Yet if the very people who are meant to act as role models behave badly, this is bound to trickle down to employees, too.

OECD analysis of foreign bribery enforcement actions reveals that most international bribes are paid by large companies, with senior management knowledge.[43] This pattern is repeated in the US, where ERC's National Business Ethics Survey 2013 found that over half of misconduct incidents involved supervisory to top management (see Figure 2, overleaf). Senior managers were responsible for a quarter of observed misdeeds and were more likely than lower-level managers to flout rules.[44]

FIGURE 2. Most misconduct committed by managers.

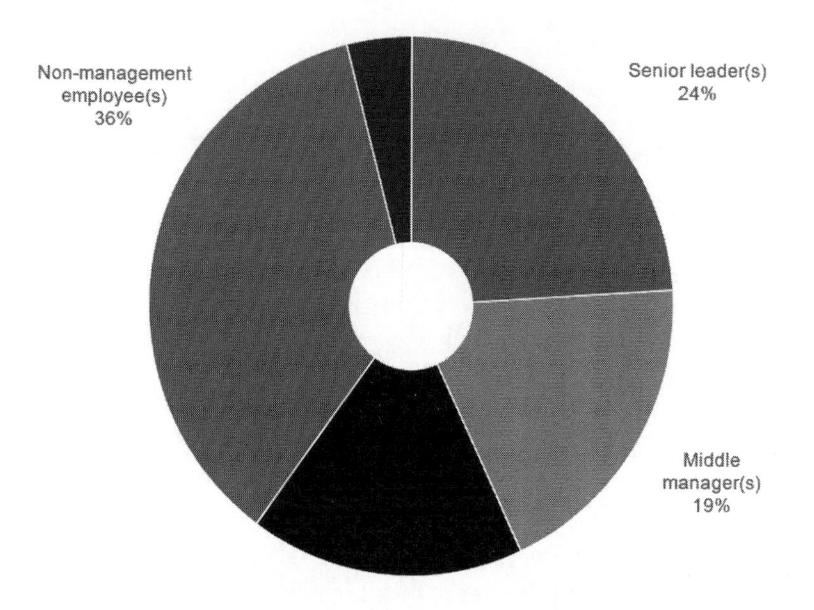

SOURCE: Ethics Resource Center, 2014. National Business Ethics Survey of the US Workforce (NBES 2013), Arlington: ERC.

Actions say it all

Integrity is a fundamental leadership attribute and it's essential for a strong, ethical culture that good conduct starts at the top. 'Do as I say, not as I do' cannot be the basis for a culture of integrity. Ethical leadership includes the following traits:[45]

- Aligning thoughts, words and deeds.
- Modelling the behaviour we ask of others.
- Learning as well as teaching.
- Considering stakeholder needs, including global society and the planet.

Leaders who demonstrate 24/7 integrity and establish ethical conduct as a priority by putting in place high standards, setting a good example and communicating openly will exert the positive influence on employees that is the oxygen of strong ethical culture. Follow-through is vital. A good example, according to Ethisphere's Timothy Erblich, is GE. 'When someone raises their hand they'll get a call from GE President and CEO Jeff Immelt or someone to say "Good job, we're glad you did that!"'[46]

Making tough calls

Modelling good ethical behaviour isn't always easy: it involves having to make tough choices and trade-offs. For example, if you're going to adopt a 'zero tolerance' approach to bribery and corruption, you have to be willing to stick to your guns and forgo tainted business opportunities to stay clean, as Alcoa did in Russia (see Box 6).

BOX 6: Zero tolerance of bribery[47]

Bill O'Rourke from the Wheatley Institution, and former Vice President, Sustainability and Environment, Health and Safety at Alcoa, knows how challenging taking a tough ethical stance can be from his time as President, Alcoa Russia. 'It wasn't easy. Russia has an entrenched system of corruption – that quickly became apparent – but consistently I said: 'We don't accept or pay bribes. We don't condone it, we don't participate in it, and we just won't do it.' It cost us as a company. When you don't pay in Russia the supply chain stops working. I had an annual capital budget of $100 million and I only spent $20 million in the first year – we couldn't get our investments underway.'

It cost O'Rourke personally, too – he was under pressure to get the investments underway and his bonus was tied to implementing them in the first year – 'But I didn't do it. We were delayed a year which wasn't good, but eventually we got there.'

Remaining steadfast under pressure is a core attribute of integrity. So too is knowing when to act decisively, something Aflac CEO Dan Amos had to do in 2011 when a brand issue arose following the tsunami in Japan (see Box 7).

BOX 7: Where the rubber meets the road[48]

It's not enough to talk about ethics, you have to demonstrate it, says Aflac's Audrey Tillman. Here's what Aflac's CEO did.

'In 2011 when a terrible earthquake and tsunami struck Japan where we do 75% of our business, we were the first company to contribute to the emergency funds. In spite of recurring aftershocks, Dan Amos immediately set off to demonstrate our concern for the affected people and ensure them their claims would be processed promptly and with great care.

'At the same time, our celebrity spokesperson, the voice of the Aflac Duck, tweeted some unpleasant jokes about the tragedy. Yet in spite of the fact that every television commercial we had prepared contained his voice, Dan Amos fired him within one hour of hearing about the tweets, knowing it would put our entire marketing campaign at risk.

'In this case, our ethical reputation outweighed our marketing campaign. Dan acted the way that he did because it was the right thing to do. The general public in both the US and Japan responded to his swift, decisive and appropriate action: Aflac received a barrage of accolades and an estimated $39 million worth of positive news stories. While that is not why we did what we did, it's nice to be rewarded for doing the right thing.'

Tips for dealing with tough calls include:

- Making your policies on unethical practices clear to business partners and sticking to them resolutely, even if it means a short-term loss. The value of good reputation is worth more in the long run. Brief the board too – tough calls are easier with their support.
- Giving employees permission to make tough calls and let them know they have your full support. When misconduct occurs, act quickly and transparently to put things right. Be willing to fire people if necessary, at every level.

Ethics is everybody's business, but the CEO, board and management have special responsibilities when it comes to creating a culture of integrity.

The role of the CEO

As the head of the company, the CEO has an oversize role in shaping the ethical culture: they set the example. The way they act, the messages they send and the objectives they choose are key determinants of company culture. Scania's Andreas Follér agrees. 'The CEO is the company embodied', he stresses, 'I can't overemphasise how crucial it is that the CEO is active. That's more or less their top task – to safeguard and remind the organisation of its culture.'[49]

The boss is a powerful influence when it comes to ethical culture change. Their role includes:

- Framing the big picture around ethics and leading the senior management team in determining the organisation's values.
- Articulating clear demands and expectations for 'how' as well as 'what' business objectives must be achieved in line with those values.
- Keeping an open door for dialogue and continually reinforcing ethical culture by being a 'storyteller'.
- Creating a positive legacy by empowering others to make right choices for the long term.

It takes around five years to push ethical culture change down through middle management. With the average tenure of a listed company chief executive just five years, their focus should be on leaving a positive legacy by embedding values for the long term and empowering others to carry on the baton.[50]

The role of the board

The board's primary function in creating and maintaining a culture of integrity is to oversee the long-term interests of the company and its stakeholders and see that value is generated in an ethical way. Its responsibilities include helping to steer corporate values and ensuring that the executive team adequately balances corporate objectives with risk management and values-led behaviour so that long-term value generation is safeguarded for all stakeholders.

A well-functioning board holds the CEO and senior leadership to account by asking the right questions, verifying that adequate checks and balances are in place to manage risk, supporting tough calls and – if necessary – changing the team if they fail to deliver against company values and stakeholder expectations. Betsy Rafael, a director at Autodesk and GoDaddy, calls this a 'noses in but hands off' tactic. The board needs to stay alert to red flags like inconsistencies, decisions that clash with values, and make sure that particularly high-stakes situations where values may be compromised 'pass the sniff test'.[51]

By working closely with the relevant steering group, internal audit, ethics and HR functions, the board can monitor the ethical climate of the organisation and health of the E&C programme. A good way to take the ethical pulse is to invite open-ended discussion about problems and use visits to unofficially 'kick the tyres' and 'get under the hood' of the E&C programme.[52]

The role of the manager

It's when values are lived consistently by every person in the company that a culture of integrity is created. Managers are responsible for embedding values through the ranks. Says RBS's Laing: 'Tone from the top is very important but not helpful if that just turns into a diktat about how you must behave. People also have to think for themselves.'[53]

Managers are key to ensuring this happens. They serve as an essential conduit to deliver and reinforce the message in a multitude of ways to frontline employees, and have the best view and insight into real-life operational challenges that people face on the job.

> ***"Tone from the top is fine, but what about the "muddle in the middle?" 30,000 of our 42,000 people are either blue collar or frontline. If you don't embed the culture in these people you've failed. They won't breach the bribery act in a material way, the Serious Fraud Office won't be knocking on your door, but if you don't deal with the culture here, the culture won't be right in the organisation, and things will become problematic."***
>
> SAM AL JAYOUSI, GROUP COMPLIANCE MANAGER, CARILLION[54]

As well as being a role model, their first job is to engage their team or unit in defining how the values contained in the CoC translate and apply in daily work. This means using their unique understanding of each role – and the challenges and risks that go with it – to develop clear guidelines. These will differ according to function: sales, for example, face very different sets of issues to R&D and this should be factored into guidelines.

Their second task is to set balanced key performance indicators (KPIs) that reward behaviour consistent with the company's values and don't put staff under unfair pressure to cut corners. Giving immediate feedback – both good and bad – is essential, along with making sure promotions reflect good performance on values and ethics metrics as well as bottom line results.

Finally, the manager needs to foster a 'speak-up' culture by making it clear that their door is always open for discussion, that reports will be acted on, and that no sanctions will be taken against whistleblowers. Providing regular feedback on investigations helps build this trust.

Ultimately, values are everybody's business. The integrity of an organisation boils down to the sum of individual choices and actions of every employee. Along with modelling 'right' behaviours, senior leaders need to ensure that effective education and incentives are there to empower each individual to do things right.

Leadership tips

- Make high personal integrity, good character and strong alignment with your company's values key criteria for promotion to all leadership roles.
- Bake the requirement to consistently walk the talk into management job descriptions and monitor whether their stated business objectives actively support ethical conduct throughout the organisation.

Tips for the CEO

- Keep ethics high on your CEO's radar with excellent regular briefings, strong messaging and great stories to tell that show why 'how' business is done is just as important as 'what' gets done.

- Encourage the CEO to take personal interest in, and openly celebrate, great examples of ethical employee behaviour and work with them to build ownership for legacy values into management systems so culture takes hold for the long term.

Tips for the board

- Consider appointing a non-executive director with expertise in ethics or consult outside expertise to stay abreast of good practice and societal expectations. Engage publicly in efforts to create industry frameworks and good practice.

- Check that the E&C function is appropriately empowered. Keep an open door for discussion of dilemmas and trade-offs and make your support for long-term, ethical value-creation over short-term, unethical profit widely known. Be prepared to fire the leadership team if they fail to deliver.

Tips for managers

- Engage your team in tailoring ethical guidelines to each role and serve as a conduit to push messages down and send feedback up the tree from the front lines.

- Set balanced KPIs, avoid undue pressure, give immediate feedback, and foster a speak-up culture by keeping your door open, protecting whistleblowers and always reporting back.

CHAPTER 6

Educate: Training and Awareness

IT IS THE ATTITUDES AND DAILY ACTIONS of every individual in the business that add up to corporate culture. To create a culture of integrity, tone from the top is crucial but not sufficient - raising awareness and ensuring that personal and organisational values are aligned are also vital. 'Saying "we have tone from the top and zero tolerance" means nothing to anybody unless you're specific about what that means', affirms TeliaSonera's Michaela Ahlberg. 'Change doesn't happen just by giving information - people need to feel it comes from them. My job is to make 27,000 people feel like "this is my idea".'[55]

To create a culture of integrity you need everyone's buy-in at all levels to ensure the whole organisation is onboard. Effective E&C training tells people what is required, raises awareness about why it matters and empowers them to make the right choices by showing them how. This is the aim of the Ethics at Electrolux programme (see Box 8).

BOX 8: Ethics at Electrolux[56]

Rollout of the programme began in 2011 to inform employees of company expectations for personal and corporate ethical accountability. The Electrolux Group Code of Ethics provides an overview of expected conduct, supported by several additional policies and guidelines such as the Code of Conduct and Environmental Policy. Available in 12 languages, the ethics programme translates these policies into action and practices via a guidance booklet, internal web portal with additional policy information, films and training-learning modules. The introductory e-learning module presents the company's expectations for ethical conduct. It uses workplace dilemmas to highlight relevant company policy and appropriate courses of action including:

- An overview of the company's vision, values and the Electrolux Foundation – the Group's principle of conduct covering three areas: Respect and Diversity; Ethics and Integrity; and Safety and Environment – which together govern how stakeholders should be treated and how business is done.[57]
- The procedure for reporting concerns, including use of the anonymous third-party ethics hotline and whistleblower protection assurance.
- Short dilemma scenarios on topics such as bribery and corruption, misuse of company assets, anti-trust, conflicts of interest, environmental protection, discrimination and harassment and what to do in each case.
- Links to more information via the company's Ethics Portal.

What is required

Employee integrity 101 starts with the 'what'. Your core values, regulatory obligations and any other industry or voluntary standards you are committed to upholding must be crystal clear. Together, your values statement, CoC and supporting policies and guidelines provide the reference framework for ethical conduct. Everyone in the organisation needs to be fully versed in the major risks for your sector, what you stand for as a company, how you aspire to treat others and what society expects of you. This creates the baseline from which all other E&C awareness initiatives can build.

A frequently used format for rolling out baseline expectations is mandatory ethics training, either face-to-face or online. Getting everyone in the organisation to do this, as global healthcare company Novartis does, establishes ethics as a priority and ensures the whole group is on the same page.

> ***"All Novartis Group company employees are required to complete integrity and compliance training. Training and communication help to maintain a culture of ethics and improve compliance, particularly when supported by an infrastructure that provides guidance and advice. Training is offered in a range of formats, including e-training, face-to-face programmes and workshops. E-training courses exist in 17 languages highlighting topics of our Novartis Code of Conduct."*** NOVARTIS[58]

As well as compulsory baseline training, the most effective ethics awareness programmes use many different channels and formats to reinforce the message, for example:

- Interactive online modules
- Classroom-based modules
- Group discussion
- Leadership roadshows
- Video and animation
- Mobile apps
- Print

Scania's *Doing things right* booklet is publicly available on its website as a downloadable pdf, setting out how to deal with common ethical dilemmas (see Box 9).

BOX 9: *Doing things right*[59]

At Scania, ethics and integrity are the responsibility of each and every individual. *Doing things right* was developed as a guide to spell out that commitment and underline the company's expectations for ethical business conduct. Designed in a handy A5 Q&A format, the Scania ethics booklet has been translated into six languages (with more in the pipeline) and is widely available to employees, dealers and distributors. The booklet:

- Describes the Scania way of doing business, with reference to key policies and guidelines and the company's core values.

- Presents a handy tool – the ethics test – made up of a series of six questions with yes or no answers, including what to do in each case.
- Using scenarios walks people through the right course of action when they encounter ambiguous situations or ethical challenges – so-called 'grey areas' – in the course of doing business.
- Provides links to further information and explains how to raise a concern.
- Invites dialogue and encourages people to share examples from their own experience.
- Outlines the consequences for failing to meet Scania's ethical standards.

Why it matters

As well as setting out expectations, providing the rationale for ethical behaviour is vital. It may seem obvious to you why 'doing the right thing' matters, but don't assume everyone sees the world that way. By joining the dots between individual actions and issues like corporate risk, reputation, brand value and stakeholder trust, you help people to see how their daily choices impact others and affect company value. Making that 'big picture' connection holds people responsible and accountable for their actions.

This tallies with experience at RBS. 'Where we've been more successful', confirms Sefton Laing, 'is where we've linked ethics to overall objectives of the bank and explained why doing the things we're saying helps – in our case being number one for trust, advocacy and personal service. You give people a good reason why you're encouraging them to behave in a personal way.'[60] RBS has introduced a tool called The YES check, which guides employees through the decision-making process, highlighting the connection between a potential action and its wider organisational implications (see Box 10). IBE has a similar, free tool that can be customised to suit, available to download from **http://www.saynotoolkit.net**

BOX 10: The YES check[61]

'Our customers, colleagues and the communities in which we do business trust each of us to be thoughtful and professional in everything we do. They expect each of us to exercise good judgement and to do the right thing. We use our values to help think through decisions and make sure we do the right thing. When in doubt, we use the YES check for guidance. Decisions are not always straightforward. The YES check can help us. It's a tool, not a rule.'[62]

Ask yourself . . .

1. **Does what I am doing keep our customers and the bank safe and secure?** Consider the impact of what you are doing. Rehearse a briefing with your boss.
2. **Would customers and colleagues say I am acting with integrity?** Consider: would I do this to someone in my family or a friend? Would I do it to myself?

3. **Am I happy with how this would be perceived on the outside?** Consider the impact of this in the outside world. Try writing the press release – does it sound good for customers?
4. **Is what I am doing meeting the standards of conduct required?** Think. If you are unsure then seek a second opinion.
5. **In five years' time would others see this as a good way to work?** Will this have a positive impact? Imagine writing it on your CV.

How to do it

The aim of E&C awareness and training is to get co-workers up to speed with organisational standards and embed values so that considering ethics in decision-making and following correct procedure becomes automatic. It's when people are empowered, motivated and supported to do the right thing that you get a culture of integrity as an outcome.

A hallmark of effective training and communication is tailoring it to particular roles and activities so people can see explicit links between the organisation's ethical standards and their own daily activities. Context is important: people working in higher risk functions like management, sales, procurement, customer service, government affairs, finance and accounting, frontline operations in emerging markets or joint ventures face particular day-to-day pressures. Training tools should take this into account by highlighting red flags and appropriate courses of action in a readily accessible format (see Box 11, overleaf).

BOX 11: The Right Way app[63]

Be creative and play around with different tools that make it easy for your people to feel informed and stay up to speed. The Ford Motor Company, for example, developed a compliance mobile app called 'The Right Way'. Available free on iTunes, the aim is to enable Ford's global workforce and business partners to access digestible policy summaries anytime, anywhere they encounter warning signals to help them make the right decisions.

Even within functions, junior and senior level staff encounter different sorts of ethical dilemmas, and training content and format should be adjusted accordingly. For example, research indicates that formal ethics training is more effective among junior employees, while informal dialogue or role-play may be better for helping senior managers develop the necessary skill-set to deal with complex 'grey' areas.[64]

A good way to do this is to use stories, case studies and scenarios that show company values in action. The IBE has a collection of several hundred short scenarios that individual organisations can tailor as needed. They recommend weekly sessions at management meetings, for example, where you pull out a scenario, turn to your neighbour, discuss it for five minutes and stop. By doing this kind of thing regularly – continuously – it 'gets into the culture' in the same way as Health and Safety. A strongpoint of this approach is that peers work on ethical dilemmas as a team which builds shared habits, a vital influence on behaviour.

“At Alcoa, we would send them a ten-question survey once every four months. The questions weren't hard but it was more to remind them things like, 'When offered a bribe, what would you do?'” Bill O'Rourke, former President, Alcoa Russia[65]

Encouraging a speak-up culture

One element of training and building awareness that deserves special mention is convincing individuals to speak up when they encounter misconduct. Creating a culture of openness where people feel they have permission to voice concerns without fear of reprisal is a cornerstone of organisational integrity.

'The most effective training and communicating is through authenticity and the best way to establish that is through two-way communication', says Aflac's Audrey Tillman. The company has put in place a mechanism, the Aflac Trust, which allows people to express concerns while protecting their anonymity, building confidence that concerns will be taken seriously and dealt with appropriately.[66]

Common ways to establish a speak-up culture include:

- Open dialogue and tough call chats – keeping an open door to discuss tricky ethical decisions.
- Clause in the CoC guaranteeing whistleblowers support and freedom from retaliation.
- Third party hotline or whistleblower mechanism for reporting values and compliance violations.
- Reporting back to employees on follow-up actions.

- Mentoring and support.
- Celebrating employees who speak up.
- Rewarding those who speak up with positive performance appraisals.

Particularly as misconduct often involves a superior (see Figure 2 on page 44), it can take great personal bravery to stand up for your values and speak out, and one of the best ways to build this up is to practise. This is the foundation of the Giving Voice to Values (GVV) programme developed by Mary C. Gentile. GVV is based on the idea that rehearsing the behaviours we'd like to have and building up 'moral muscle memory' can help us 'pre-script' ourselves to act ethically, so that we're more likely to act in line with our values when challenged.[67]

As well as increasing employee awareness, encouraging an open, speak-up culture and rehearsing desired behaviours often to build good habits, a prime driver of conduct is how you reward and recognise people. We look at this next.

Five tips on training

- Make baseline expectation training compulsory for everyone and address the what, why and how. Tailor relevant material and channels according to different functions and levels.
- Take out the jargon, make it short and simple, but be specific about what you mean. Mix it up: avoid eLearning fatigue by using a variety of techniques: online, face-to-face, scenarios and role-play, mobile apps and tools, formal and informal. Make it fun!

- Repeat it often and revisit it regularly – piggyback on messaging you've already embedded successfully, like Health & Safety or Quality – and practise ethical responses to build up 'moral muscle memory'.
- Communication is a two-way process – be open and use dialogue about ethics to take the ethical pulse of the organisation. Make it clear that speaking up is valued and won't be penalised.
- Keep your programme well staffed and resourced and build in metrics and measurable KPIs to capture changes in attitude and programme impact.

CHAPTER 7

Reward: Performance and Incentives

YOU'VE NO DOUBT HEARD THE STORY OF PAVLOV'S DOG – where the animal comes to associate hearing a bell with getting food and learns a new behaviour. Human beings are not so different: if we connect particular actions with rewards we adapt our behaviour to suit. A very powerful way to influence employees is by tying pay and promotion to strong ethical conduct.

Incentivising ethical behaviour through positive support is highlighted by both Transparency International and the OECD as an important means to demonstrate your organisation's E&C commitment.[68] This means working cross-functionally with HR to bring compliance and compensation together so that ethics becomes integrated into performance evaluations. One company that has joined the dots between a culture of integrity and annual performance reviews is Novartis, which has made good conduct a component in its associates' yearly assessments, with the explicit aim of incentivising ethical behaviour.[69]

> ***"Performance Management plays an important role in embedding ethical values and behaviour in the culture of the organisation. It is an excellent way to reinforce desired behaviour year after year. Knowing that they will be appraised on how they***

do business as well as what they achieve will increase employees' sensitivity to the ethical matters they may confront in their day-to-day business. ❞ INSTITUTE OF BUSINESS ETHICS[70]

What gets rewarded gets repeated

The high-jinks unethical conduct rife across banking and financial services in recent years arose through a perfect storm of limited oversight, aggressive mergers, estrangement from core purpose and reward structures that drove short-term revenue-hunting and personal enrichment at any cost. In such an environment an ethical culture never stood a chance.

Bonus-driven corporate culture has been recognised by the UK Parliamentary Banking Commission as a key factor in the 2008 financial crisis.[71] More recently, in Australia, investigations into financial planning malpractice at National Australia Bank have revealed remuneration structures heavily biased towards product sales and new clients rather than creating value for existing customers.[72] Sending better signals to employees via recalibrated reward systems that promote individual accountability and strike a balance between earning revenue and responsible, client-centred risk management, would go a long way to prevent this. Hardwiring ethics to incentives is among the best resources in the corporate toolkit.

Handelsbanken is an interesting example in this regard. Unlike many in its sector, the Swedish full-service bank has a long-term, low-risk tolerance, sustainable value creation strategy with a business model centred on customer satisfaction. In 2014, the Group's operating profit grew by 6% to SEK 19,212m – the highest figure in the Bank's history – while achieving above sector-average customer satisfaction. Salary, pension systems and profit-sharing are explicitly recognised by the bank as ways of boosting corporate culture. Two cornerstones of the Handelsbanken way are its variable remuneration policy and profit-sharing scheme, *Oktogonen*, which are deliberately designed to encourage responsible conduct, long-term value creation and employee retention (see **Boxes 12** and **13**).

BOX 12: Tying rewards to risk[73]

At Handelsbanken, the board decides on remuneration policy. Fixed remuneration applies to more than 97% of the Group's employees and is applied 'without exception' to executive officers, staff who decide on granting of credits and employees in control functions. In 2014, variable remuneration was less than 2% of the Group's total salary costs and fees. Key features of the approach are:

- Variable remuneration is limited, requiring a special decision by the CEO with the board deciding the final amount.
- Management employees who can affect the bank's risk profile, e.g. by deciding on credit, market, liquidity, commodity, currency, or interest rate risk limits, can have only fixed remuneration.
- Only employees from units deriving their profits from low-risk transactions are entitled to variable remuneration. This must be designed to discourage unhealthy risk-taking, be within the limits of the bank's risk tolerance and strike a reasonable balance with fixed remuneration.

- It is only paid in cash and disbursement of at least 40% must be deferred at least three years if it amounts to SEK 100,000 or more. Where it exceeds senior management remuneration, 60% is deferred for four years. Payment and right of ownership only accrue after the end of the deferment period.
- It can be removed or reduced if losses, increased risks or increased expenses arise during the deferment period or in view of the bank's financial situation.
- No employee may receive variable remuneration of more than 100% of his/her fixed remuneration.

The Handelsbanken approach to long-term incentives may not work for everyone, but it highlights how remuneration can be used to shape organisational culture and achieve corporate goals. A variation is to introduce discretionary variable-pay awards for outstanding ethical leadership.

BOX 13: Tying rewards to loyalty[74]

Instead of short-term bonus systems, Handelsbanken applies an employee profit-sharing scheme called *Oktogonen*, managed by a foundation. Allocations are made if the bank achieves its financial goal of better-than-average profitability, which has been the case every year since 1973, bar two. One third of the extra profits can be allocated to employees, with the amount limited to 10% of shareholder dividends. If these are reduced no allocation is made to the foundation.

This helps to align employees' interests with corporate goals, with the aim of making cost-awareness and caution an intrinsic part of Handelsbanken's corporate culture. All employees receive an equal part of the allocated amount, regardless of position and work tasks.

The long-term profit-sharing scheme covers 98% of the Group's employees, with payments made in the employee's 60th year. Funds are invested in Handelsbanken shares, making *Oktogonen* one of the bank's largest shareholders. *Oktogonen* has two representatives on the Handelsbanken board, giving employees a say at board level.

Pressure to perform

Pay and performance go hand-in-hand. Relaxing expectations by setting more sustainable goals for long-term stakeholder value creation will help create a climate in which integrity can flourish.

Interestingly, a study by the Economist Intelligence Unit found that between 2010 and 2013 nearly half (43%) of financial services companies surveyed had introduced career or financial incentives to encourage adherence to ethical standards. Yet over the same period, more than half of respondents (53%) thought career progression would be difficult 'without being flexible on ethical standards', a figure that rose to 71% for investment bankers.[75]

This shows the significant challenge ahead for this and other sectors in changing industry norms and highlights the importance of revisiting corporate objectives and expectations as well as incentive structures. If managers set unrealistic or overly aggressive performance targets that put people under pressure to deliver, the likelihood is that risks get taken or corners cut to meet them, creating what anti-bribery advisor Richard Bistrong calls a 'zero-sum game' between performance and compliance where ethical behaviour risks being seen as 'bonus prevention'.[76]

Setting balanced KPIs has a key part to play in creating an ethical culture. Bistrong recommends 'taking a look under the hood of bonus plans' for high-risk personnel and encouraging open discussion of frontline risks. The resulting E&C insight can be used to adjust business strategy, forecasts and incentives to ensure they work with – not against – your company's ethical standards.

> ***"How many companies, including C-Suite personnel and those who serve on the relevant board committees, are looking at the business strategy itself as a potential red flag of corruption? I call attention to business strategies and growth forecasts in regions that have reputations for corruption and ask if anyone is asking when those targets are achieved "how did we get there?" Or is it all high-fives?"*** RICHARD BISTRONG, ANTI-BRIBERY ADVISOR[77]

Even simple things like stretching decision-making timeframes can make a difference: RBS has built in time for senior leaders to look at issues in greater depth rather than trying to action everything straight away. Research shows people are more likely to display unethical behaviour at particular times of day when it's harder to resist temptations, so timing important meetings or decisions for when people are at the top of their game could also help.[78]

Penalties, awards and recognition

If you say you have zero tolerance for ethical breaches you have to follow through – the standard is 100% compliance and you need to show consistent application of penalties and consequences when violations occur – including pressing charges where necessary. This is an essential part of walking the talk and modelling high ethical standards to your organisation: tolerance and compassion are important, but failure to follow through can undermine the whole system and weakens your culture of integrity.

On the flip side, informal systems like public praise and awards for good conduct reinforce culture by recognising individual valour. US insurer Aflac's 'Aflac Way Honors' is one example of this. Colleagues can nominate each other for ethical behaviours, which can lead to gifts and prizes for maintaining the company's ethical culture.

Five tips on performance and incentives

- Work cross-functionally with HR to bring together compliance and compensation so that ethics and risk are integrated into performance management. Take a look 'under the hood' of bonus plans for high-risk personnel and encourage open discussion of frontline risks.

- Use remuneration to achieve values-driven goals and harness bonus culture for 'good', e.g. by introducing discretionary variable pay awards for outstanding ethical leadership.

- Revisit corporate objectives and KPIs to ensure management isn't putting employees under unfair pressure.
- Follow through: dealing with offenders gives your ethical standards legitimacy.
- Use public praise, awards or a pat on the back for good conduct to actively reinforce integrity.

CHAPTER 8

What To Do When Things Go Wrong

MANAGING INTEGRITY IS DIFFICULT. No-one expects everything to go right all the time and indeed sooner or later, something is pretty much bound to go wrong. The trick lies in knowing what to do when it does.

The corporate pantheon is full of examples, good and bad, of how to respond when disaster strikes. Aflac responded swiftly and effectively when an ethical problem arose with the 'voice' of its iconic duck (see Chapter 5). By contrast, the painfully limp leadership response from soccer's ruling body FIFA to protracted accusations of systemic corruption is a perfect example of what not to do. But what can we learn from the companies that have got 'getting it wrong' right, so to speak?

Siemens: Crisis as opportunity

German engineering giant Siemens is the poster child for anti-corruption makeovers (see Box 14, overleaf). Following a massive conduct scandal in 2006, the company put in place what is widely considered to be the 'gold standard' of compliance programmes, praised by the OECD and US federal authorities for its comprehensiveness and rigour. The company is now held up as a benchmark for anti-corruption. Key factors in the Siemens E&C success story have been:

- Harnessing lessons learned from misconduct to drive whole-organisation culture change.
- A multi-pronged approach across structures, procedures and culture.
- Strong execution.
- Honest recognition that changing corporate culture takes time – preparing for a marathon, in other words, not a sprint.

BOX 14: 10 key lessons from a crisis

In 2006, Siemens was struck by scandal when it emerged that extensive bribes had been paid to secure contracts for power generation equipment in Italy, telecommunications infrastructure in Nigeria, and national identity cards in Argentina. Siemens identified $1.6bn in 'questionable payments' made around the globe from 2000–2006.[79] A huge international corporate bribery and corruption investigation ensued. Overall, the scandal cost Siemens €2.5bn, including €2bn of fines.[80] The company also suffered significant reputational damage, was excluded from dealing with certain clients and its employees had to endure the shame of intense public scrutiny and criticism. Importantly, Siemens proactively used the crisis, with 10 key actions to drive company-wide change:

1. Hired Peter Löscher as the first outsider CEO since the company was founded in 1847. First action: tour global operations to get the facts.
2. Established zero tolerance for bribery and corruption as the

tone from the top and declared an amnesty for employees to speak up.

3. Revamped governance including dismantling the two-tiered board system, asking 80% of the Managing Board's membership to leave, and creating a new cross-functional board including legal counsel and compliance, and supply chain and sustainability.
4. Reduced operating units from 12 to eight and reduced the company's complex matrix structure to three divisions.
5. Appointed an external advisor from Transparency International and over 500 full-time compliance officers.
6. Introduced a 'gold standard' compliance programme, based on prevention, detection and response, including clear reporting channels, strict procedures, hotlines, a risk evaluation web portal and an external ombudsman.
7. Carried out over 900 internal disciplinary actions, including dismissals.
8. Launched a comprehensive employee education and awareness programme. By 2008 over half its 400,000 global workforce had received anti-corruption training.
9. Initiated internal 'integrity dialogues' to promote open discussion of dilemmas and external reporting of compliance-related issues.
10. Launched a global Siemens Integrity Initiative that supports efforts to fight corruption and fraud.[81]

TeliaSonera: Restoring trust

Swedish telecom operator TeliaSonera's anti-corruption journey is more recent, after bribery allegations surfaced in 2012 over deals made in central Asia (see Box 15). Sweeping efforts to address the issue and restore the company's reputation have included:

- A change of senior leadership.
- Full overhaul of the E&C function.
- Updated policies and frameworks.
- Extensive employee anti-corruption training.
- A transparency report.

Like Siemens, TeliaSonera sees this as a long-term investment and has seized the opportunity to put in place a state-of-the-art risk management and E&C system, of which culture change is a core element.

BOX 15: Seven ways to rebuild trust[82]

TeliaSonera is an international telecom group with operations in 18 countries from Norway to Nepal, including a strong presence in Central Asia. In 2012, allegations surfaced that it had participated in corrupt business deals in Uzbekistan to secure frequencies and licences. After initial rejection of the claims, a review revealed the group had failed to properly assess its local partner, including possible instances of bribery. With international investigations still underway, since 2013 the company has taken significant steps to restore leadership trust and address apparent weaknesses in its

E&C framework:

1. A top-level management shake-up, including dismissal of the CEO and CFO and appointment of a new General Counsel, M&A function and new Chief Ethics and Compliance Officer with access to the board.
2. A complete revamp of the E&C framework around an eight-step cycle, based on Prevent, Detect and Investigate principles.
3. Revision of key policies on anti-corruption, corporate giving, sourcing and procurement, M&A and non-retaliation, supported by specific guidelines and instructions, with new KPIs in the pipeline.
4. Setting up a cross-functional network to support and promote ethical behaviours, including a multi-cultural E&C team at HQ and local Country E&C officers.
5. Anti-corruption and human rights awareness training for employees, including face-to-face and dilemma-based workshops, with 4691 employees completing face-to-face training in Q2 2014.
6. A Special Investigations Office, established in April 2014, with a new investigation protocol and Ethics Forum for disciplinary and management actions.
7. An internal review of Eurasia operations by the board, including full, ongoing cooperation with relevant authorities.

As well as putting effective and adequate procedures in place, a strong focus going forward is to strengthen company culture. TeliaSonera is also collaborating with others on the systemic level to lower conduct risk across the telecom sector as a whole.

Six key steps

Both examples showcase the consensus on key good practice steps when it comes to dealing with misconduct, summarised in the following table.

STEP	ACTION
1. **Recognise the problem**	Openly acknowledge that a violation has – or may have – taken place. Inform stakeholders. Issue a public apology and commit to investigate. Make it known that, if confirmed, there will be zero tolerance and things will change. Don't be secretive or defensive.
2. **Share the facts**	Don't leap to conclusions: investigate first and get the facts. Cooperate fully with authorities. Declare an amnesty if necessary to encourage honesty.
3. **Conduct an inquiry**	Identify the mechanisms and analyse how and why it happened so you can prevent it next time.
4. **Take action**	Follow through. As soon as the facts are clear, take swift and decisive remedial action. If necessary, penalise offenders. Show zero tolerance, but don't blame individual scapegoats for systemic failures.

STEP	ACTION
5. **Learn**	Harness the learning opportunity a crisis brings and share widely lessons learned. Adjust policy and procedures accordingly and use real-life cases as dilemma scenarios to raise awareness and practise 'doing the right thing'.
6. **Report**	Monitor how your E&C programme performed – was it fit for purpose? Track progress and report publicly on findings and outcomes.

Nobody's perfect

It's tempting to want to be perfect, but the ugly truth is things can and will go wrong from time to time. When they do, openness is key.

It may be comforting to know that investors aren't looking for a squeaky-clean balance sheet when it comes to ethics; in fact, they're likely to be suspicious if things sound too good to be true. Says Anna Young-Ferris, Leader UN Principles for Responsible Investment (UNPRI) Asia Pacific Academic Network, "Investors want to see consistent, well-managed environmental, social and corporate governance practices. That includes how management deals with a crisis when things go wrong. If it's only good news stories, that can imply there's something lurking beneath. It's how they manage in good times but more importantly in bad times that's the key indicator."'[83]

> ***"Bad news doesn't improve with age."*** DAN AMOS, CEO AFLAC[84]

Leading practice, according to Ethisphere's Tim Erblich, is to admit mistakes and, strange as it sounds, pray your hotline gets used. 'Receiving a lot of calls indicates that people feel they can raise their hands. With whistleblower mechanisms, it matters what you do with it: if you don't promote it, say how it's being used, share results with employees or your board, it's not very useful.'[85]

So in summary: treat every incident as an opportunity for openness and improvement, but remember prevention is your ally. Be on the lookout for common red flags to help your culture stay strong.

Red flag checklist

Here are 15 common signs that you may have an ethical problem-in-waiting:

- High-risk sector and high-stakes decisions.
- Activities in frontier markets and/or extensive use of agents and third party intermediaries.
- Involvement in public works and public procurement contracts.
- Complex organisational structure with little cross-functional integration or communication.
- Extensive supply chain or customers who do not share your values.
- Lack of a unifying corporate purpose ('one company') and core values ('our way') that put stakeholders at the heart.

- Aggressive business goals, strategies and KPIs that prioritise short-term results or new sales figures over long-term value creation.
- Weak, inward-looking leaders who don't walk the talk, are blasé about risk or view ethics as a hurdle or checklist exercise.
- Figures or outcomes that are too good to be true – success can breed complacency.
- Phrases like 'we didn't have this conversation', 'no-one needs to know' or 'just this once'.
- Under-resourced, 'bolt-on' E&C programme with no access to the board or CEO.
- Limited or one-off training and communication efforts.
- Pockets of misconduct – this can indicate a subculture or values clash.
- Failure to follow through when violations occur or acknowledge good conduct
- Lack of openness, fear or reluctance to report (a silent hotline) or a mismatch between company and stakeholder perceptions.

CHAPTER 9

Keeping Your Culture Strong

COMPANIES WITH STRONG CULTURES OF INTEGRITY have a number of things in common. They all put ethical values at the heart of corporate culture and make efforts to stay attuned to the changing universe of risks and issues impacting their business. Specifically, they:

- *Align* their values, business goals and behaviours with stakeholder needs and expectations.
- *Integrate* ethical behaviour into standards, management systems and incentives.
- *Reinforce* their culture through training and awareness, awards and communication, measurement and reporting.

It is this cycle that, over time, creates a strong culture of integrity in an organisation (see Figure 3, overleaf).

To sustain your culture, you need to harness this cycle to foster integrity at three distinct levels.[86]

At the personal level

Ethical decision-making boils down to individual choices. To create a culture that goes the distance, it's important that everyone is pulling in the same direction with your company values as their 'North Star'. A key

FIGURE 3. Culture of integrity.

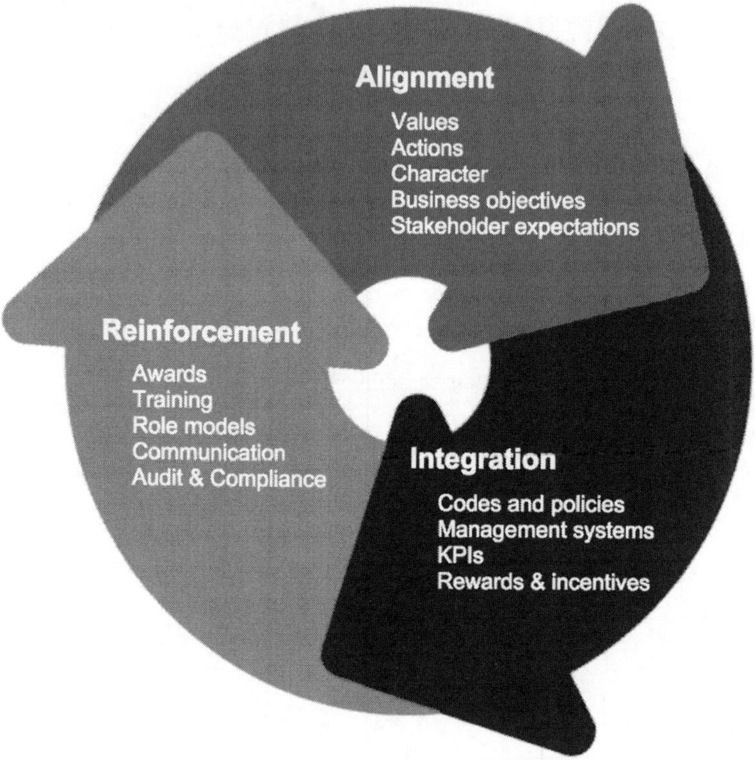

Alignment
Values
Actions
Character
Business objectives
Stakeholder expectations

Reinforcement
Awards
Training
Role models
Communication
Audit & Compliance

Integration
Codes and policies
Management systems
KPIs
Rewards & incentives

SOURCE: © One Stone Advisors 2015.

ingredient in Aflac's strong corporate culture is having everybody living it from top to bottom – no-one is 'excused' from ethics, and everyone is expected to conduct business with ethics in mind. Make personal responsibility an expectation in yours.

As well as continually reinforcing the message, motivating people to act ethically through rewards and recognition and regularly refreshing their knowledge and skills, consider adopting a customised tool that helps employees stop and think before they act, as RBS has done (see Box 10). The long-running Green Cross Code pedestrian safety campaign – Stop. Look. Listen. Think – is an enduring example of a reinforcement mechanism that builds awareness until behaviour becomes engrained.

Character is also very important. If somebody's values are completely at odds with what your organisation stands for, creating a culture of integrity is going to be a bumpy ride. Without shutting out diversity, put personal values at the centre of your hiring procedures. Probing to see if potential new recruits are culture-aligned saves time, trouble and resources in the long run. Hiring individuals who 'fit' with your mission and purpose will also lead to stronger teams, happier employees and greater engagement.[87]

> ***"We all need to crack ethical recruitment. You don't just do reference checks, you also test them on their ethical position in the interview, give them questions, dilemmas and scenarios. You should ask this very early on – the company values have to be aligned with personal values."*** SAM AL JAYOUSI, GROUP COMPLIANCE MANAGER, CARILLION[88]

At the organisational level

Creating and maintaining a culture of integrity at organisational level has been the main focus of this book. As set out in Chapters 3–7, a number of different aspects are involved and keeping your culture strong requires sustained momentum on all fronts. Pinning up a poster and hoping people remember just won't cut it.

Famous examples of strong corporate culture practice like the Ritz-Carlton Daily Line-Up shed useful light on the importance of living your values by repeating, renewing, recognising and rewarding values-based behaviour. A few minutes invested in reinforcing ethical conduct every day can go a long way to making it a life-long habit.

> ***“Daily Line-Up is the opportunity to reconnect each employee with their purpose and their mission before they start their day. Although there may be many elements to your Line-Up – such as corporate announcements and birthdays – the emphasis should be on your culture and values.”*** THE RITZ-CARLTON LEADERSHIP CENTRE[89]

All ethical cultures weaken over time, so periodic renewal is vital. You should plan to revisit your values to ensure they're fit for purpose whenever the risk landscape shifts, such as when you enter a new market, or new ethical obligations emerge and use this opportunity to refresh your commitment. At the very least experts recommend doing this every two to three years. You can dovetail this with stakeholder materiality processes to keep your values aligned with societal expectations.

To gauge the strength of your culture and spot gaps or signs of slippage make sure you put in place good metrics to measure the impact of the E&C programme. As well as being a useful way to report on effectiveness to the board, keeping track of training, whistle-blowing and misconduct figures and monitoring legal fees, employee engagement and customer satisfaction, for example, can indicate how conduct is trending, and highlight areas for improvement.

A powerful way to test your organisational culture is through third party certification such as the Investing in Integrity Charter Mark (**http://www.**

cisi.org) and rankings like Ethisphere's World's Most Ethical Companies (**http://ethisphere.com**). Rigorous self-assessment, third party assurance and good practice benchmarking all have a vital role to play in maintaining high standards of conduct and ensuring that complacency doesn't set in. Holding yourself accountable by making a public commitment, setting targets and reporting annually on performance can also help to keep your culture strong.

As highlighted in Chapter 3, an ongoing challenge is how to establish, communicate and embed a single set of values across global, multi-cultural or complex organisations. This is an evolving area, but common threads of good practice include:

- Focusing on 'one company' standards and universal values rather than national laws.
- Ensuring diversity in the E&C team by appointing local staff with insight into regional customs in each country or operational unit.
- Inviting open discussion of accepted cultural practices and 'grey areas' so that agreed boundaries can be established.
- Involving local teams in adapting training materials and scenarios and making these available in local languages.
- Developing culturally sensitive ways for people to speak up without fear of reprisal.

Ultimately, establishing shared values and business practices that carry fluidly across national borders and time zones is something no company can do alone: it calls for a community of likeminded organisations working together to change the global business system as a whole.

At the system level

It's difficult for an ethical organisation to survive in an unethical world. No company is an island and reputations can get tarnished by association if somebody in your value chain does the wrong thing. When it comes to 'non-captive' parts of your business – mergers, acquisitions, joint ventures, third party agents, suppliers and even host governments and customers – thorough due diligence and shared values are the watchwords. If any of these relationships presents too high a risk, walk away.

A good way to keep your culture strong over the long term is to join a network of organisations that share your commitment. It is far easier to take a strong stand on poor practices as a group: Janaagraha's *I Paid A Bribe* (**http://www.ipaidabribe.com**) is an interesting example of how collective citizen action in India is being leveraged to drag corruption into the open and improve governance. Companies can do the same. The World Economic Forum's Partnership Against Corruption Initiative (PACI) shows how business is increasingly uniting in support of strong values.

By adding your voice to the mounting chorus for tough measures against corruption you are sending the message that unethical business is no longer acceptable. It's the right thing to do – real markets for sustainable prosperity depend on it. Increasingly, too, integrity is becoming an entry ticket for growth.[90]

> ***“Business today operates in a global market where focus on these issues is increasing and ethical business is an integral part of business practice. Companies need to have done their homework. They need to give a true picture of the challenges they face and how they are trying to manage them. That's why***

***transparency is one of our guiding principles for measuring if companies are at the forefront.*”** **HELENA HAGBERG, SPECIAL ADVISOR SUSTAINABLE BUSINESS, SWEDISH MINISTRY OF FINANCE**[91]

It also makes good business sense. The clock is ticking for those that tolerate 'ethical drift', warns Mark Carney, Governor of the Bank of England, and companies must become accountable. 'For the best in the industry, this won't be new. This is just how you run your business. But for others, who free-ride on your reputations: the Age of Irresponsibility is over.'[92]

To make sure your company is fit for the 21st century, we'd like to leave you with eight golden rules that will keep your culture on track.

Eight golden rules

1. Align, integrate and reinforce integrity at individual, organisational and system levels to create a virtuous cycle.
2. Make personal responsibility for integrity a daily expectation and deploy tools that engrain ethical decision-making as a habit.
3. Hire for character and values and reinforce good conduct through rewards, recognition and regular updates.
4. Revisit your values every two or three years in light of new risks and issues impacting your business and renew your commitment.
5. Apply metrics to measure progress and highlight areas for

improvement. Consider investing in third-party certification and reporting publicly to hold your culture accountable.

6. Take a 'one company' approach to values, but empower local teams to embed these in culturally sensitive ways.
7. Conduct due diligence on 'non-captive' parts of the business and if the risks are too high, walk away. Do business with others who share your standards to build long-term value and lower risk.
8. Join multi-stakeholder initiatives to work against corruption, foster a level playing field and turn integrity into a business enabler.

Notes and References

1. Leahy, J. 2015. Petrobras probe looks to dig up corpse. *FT.com*. **http://www.ft.com/intl/cms/s/0/4badd072-ff3d-11e4-84b2-00144feabdc0.html#axzz3d5XSIPQP**

2. OECD. 2014. *The rationale for fighting corruption*. **http://www.oecd.org/cleangovbiz/resources.htm**

3. Phipps, S. 2015. Banks still mired over conduct risk. *Ethical Corporation*. **http://www.ethicalcorp.com/stakeholder-engagement/banks-still-mired-red-over-conduct-risk**

4. Economist Intelligence Unit. 2013. *A crisis of culture: valuing ethics and knowledge in financial services* (London: EIU), p. 11. **http://www.economistinsights.com/sites/default/files/LON%20-%20SM%20-%20CFA%20WEB.pdf**

5. Kitroeff, N. 2014. Wall Street has a conscience. This professor is determined to find it. *Bloomberg Business*. **http://www.businessweek.com/articles/2014-09-02/a-professor-campaigns-to-make-wall-street-more-ethical#r=nav-r-story**

6. UN Global Compact. 2014. **http://www.unglobalcompact.org.au/news-and-events/events/corporate-panel-anti-corruption-outcomes-from-the-g20-and-b20/**

7. OECD (op. cit.).

8. Kemp, H. 2014. The cost of corruption is a serious challenge for companies. *Guardian Sustainable Business*. http://www.theguardian.com/sustainable-business/corruption-bribery-cost-serious-challenge-business

9. Carney, M. 2015. *Building real markets for the good of the people*. Bank of England speech, Lord Mayor's Banquet for Bankers and Merchants of the City

of London at the Mansion House, London. **http://www.bankofengland.co.uk/publications/Documents/speeches/2015/speech821.pdf**

10. One Stone Advisors. 2015. Interview with Sam Al Jayousi, Group Compliance Manager, Carillion plc.

11. SPIEGEL interview with Siemens CEO. 2013. *We need to see calm restored.* **http://www.spiegel.de/international/business/spiegel-interview-with-siemens-ceo-joe-kaeser-a-915314.html**

12. Adapted from Bonime-Blanc, A. 2014. How boards can gauge the effectiveness of ethics programmes. *Ethical Corporation.* **http://www.ethicalcorp.com/business-strategy/how-boards-can-gauge-effectiveness-ethics-programmes**

13. US Department of Justice and US Securities and Exchange Commission. 2012. *A resource guide to the US Foreign Corrupt Practices Act.* **http://www.justice.gov/criminal/fraud/fcpa/guidance/guide.pdf**

14. Solomon, M. 2014. 9 leadership steps for corporate culture change. *Forbes.* **http://www.forbes.com/sites/micahsolomon/2014/09/27/a-leadership-checklist-for-culture-change-and-customer-experience-excellence/** and http://www.micahsolomon.com

15. OECD. 2014. *Foreign Bribery Report: An Analysis of the Crime of Bribery of Foreign Public Officials* (Paris: OECD Publishing).

16. Roe, M. 2014. The Fed's culture war. *Project Syndicate.* **http://www.project-syndicate.org/print/fed-suggests-regulating-bankers-by-mark-roe-2014-11**

17. One Stone Advisors. 2015. Interview with Sefton Laing, Head of Sustainability Services, RBS.

18. Ethics Resource Center, 2014. *National Business Ethics Survey of the US Workforce* (Arlington, VA: ERC).

19. One Stone Advisors. 2015. Interview with Andreas Follér, Sustainability Manager, Scania.

20. Evans, J. 2014. *What Quaker companies can teach us about wellbeing at work.* **http://philosophyforlife.org/what-quakers-can-teach-us-about-well-being-at-work/**

21. One Stone Advisors. 2015. Interview with Simon Webley, Research Director, Institute of Business Ethics.

22. One Stone Advisors. 2015. Interview with Kirk Hanson, Executive Director, Markkula Center for Applied Ethics, Santa Clara University.

23. Johnson & Johnson. 2015. **http://www.jnjmedical.com.au/about/our-credo**

24. Peterson, J. 2014. Is integrity the secret to great leadership? *Forum:Blog.* **https://agenda.weforum.org/2014/09/integrity-leadership-consistency-actions-words/**

25. Bradshaw, K. 2014. Business Ethics discusses how to go about embedding an ethical culture in your organisation. *Communicating Ethical Values.* **http://www.ioic.org.uk/news/2014/october-2014/communicating-ethical-values.html**

26. Labaton Sucharow. 2015. **http://www.labaton.com/en/about/press/Historic-Survey-of-Financial-Services-Professionals-Reveals-Widespread-Disregard-for-Ethics-Alarming-Use-of-Secrecy-Policies-to-Silence-Employees.cfm**

27. Ydstie, J. 2015. Pressure to act unethically looms over Wall Street, survey finds. *NPR.* **http://www.npr.org/2015/05/19/408010692/pressure-to-act-unethically-looms-over-wall-street-survey-finds**

28. Adapted from Ethics Resource Center. 2009. *Ten things you can do to avoid being the next Enron.* **http://www.ethics.org/resource/ten-things-you-can-do-avoid-being-next-enron**

29. Aflac. *2013 Aflac Corporate Citizenship Report.* **http://web28.streamhoster.com/aflac/flipbook/2013aflacccr/html/index.html#3/z**

30. One Stone Advisors. 2015. Interview with Audrey Tillman, Executive Vice-President and General Counsel, Aflac.

31. One Stone Advisors. 2015. Interview with Timothy Erblich, CEO, Ethisphere.

32. Institute of Business Ethics. 2014. Is it possible for business behaviour to be changed from within? *IBE Press Release.* **https://www.ibe.org.uk/userassets/pressreleases/eandcpractitioners.pdf**; Coffey, F. 2014. *The Role and Effectiveness of Ethics and Compliance Practitioners* (London: IBE).

33. One Stone Advisors. 2015. Interview with Michaela Ahlberg, Chief Ethics and Compliance Officer, TeliaSonera.

34. One Stone Advisors (op. cit.).

35. One Stone Advisors (op. cit.).

36. One Stone Advisors (op. cit.).

37. One Stone Advisors. 2015. Interview with Malin Ekefalk, Director of Corporate Social Responsibility, Electrolux.

38. One Stone Advisors. 2015. Interview with Bill O'Rourke, former President, Alcoa Russia, Ethics Fellow at the Wheatley Institution, Brigham Young University.

39. One Stone Advisors (op. cit.).

40. One Stone Advisors (op. cit.).

41. Price, M. 2013. Imagination working with integrity: How General Electric creates a global culture of ethics. *Ethisphere Magazine.* **http://ethisphere.com/magazine-articles/imagination-working-with-integrity/**

42. One Stone Advisors (op. cit.).

43. OECD. 2014 (op. cit.).

44. Ethics Resource Center. 2014. *National Business Ethics Survey of the US Workforce 2013.* **http://www.ethics.org/downloads/2013NBESFinalWeb.pdf**

45. Fisher Thornton, L. 2015. **http://www.leadingincontext.com**

46. One Stone Advisors (op. cit.).

47. One Stone Advisors (op. cit.).

48. One Stone Advisors (op. cit.).

49. One Stone Advisors (op. cit.).

50. Howard, S. 2014. CEOs: the challenge of creating a lasting legacy of sustainability. *Guardian Sustainable Business, Social impact hub.* **http://www.theguardian.com/sustainable-business/2014/oct/02/ceos-challenge-creating-lasting-legacy-sustainability-short-term/print**

51. O'Toole, J. *Role of the Board in Creating an Ethical Culture*. Markkula Center interview with Betsy Rafael, Autodesk. **https://www.youtube.com/watch?v=kOm8SC8ql4w**

52. Bonime-Blanc, A. 2014. How boards can gauge the effectiveness of ethics programmes. *Ethical Corporation.* **http://www.ethicalcorp.com/business-strategy/how-boards-can-gauge-effectiveness-ethics-programmes**

53. One Stone Advisors (op. cit.).

54. One Stone Advisors (op. cit.).

55. One Stone Advisors (op. cit.).

56. Electrolux. 2014. *Ethics at Electrolux eLearning Module* (Stockholm: Electrolux Group).

57. Electrolux. 2015. **http://www.electroluxgroup.com/en/topic/the-electrolux-foundation/**

58. Novartis. 2015. **https://www.novartis.com/about-us/corporate-responsibility/our-actions/ethics-compliance/standards-integrity**

59. Scania. 2012. *Doing Things Right: A Guide to Ethics at Scania* (Södertälje: Scania Public and Environmental Affairs). **http://www.scania.com/sustainability/archive/how-scania-works/governance-for-accountability/doing-things-right/index.aspx**

60. One Stone Advisors (op. cit.).

61. RBS. 2015. *Code of Conduct.* **http://www.rbs.com/content/dam/rbs/Documents/about/code-of-conduct.pdf**

62. One Stone Advisors (op. cit.).

63. **https://itunes.apple.com/au/app/the-right-way/id950707840?mt=8**

64. Seligson Levanon, A. and Choi, L. 2006. *Critical Elements of an Organisational Ethical Culture* (Washington, DC: Ethics Resource Center).

65. One Stone Advisors (op. cit.).

66. ne Stone Advisors (op. cit.).

67. Gentile, M.C. *Giving Voice to Values.* **http://www.givingvoicetovaluesthebook.com**

68. Transparency International (TI). 2009. *Business Principles for Countering Bribery,* 1st edn 2003; 2nd ed. 2009 (light revisions); OECD. 2009. *Good Practice Guidance on Internal Controls, Ethics and Compliance* (Paris: OECD), A.9.

69. Novartis (op. cit.).

70. Steinholz, R. and Dando, N. 2014. *Performance Management for an Ethical Culture: An IBE Good Practice Guide* (London: IBE).

71. Phipps, S. 2015. Banks still mired over conduct risk. *Ethical Corporation.* **http://www.ethicalcorp.com/stakeholder-engagement/banks-still-mired-red-over-conduct-risk**

72. Ferguson, A. 2015. National Australia Bank will need more than remorse to satisfy Senate grilling. *Sydney Morning Herald*, 3rd May 2015.

73. Adapted from Handelsbanken. 2015. **http://www.handelsbanken.se/shb/inet/icentsv.nsf/vlookuppics/investor_relations_en_hb_14_highlights/$file/hb_14_highlights.pdf**

74. Adapted from Handelsbanken (op. cit.).

75. Economist Intelligence Unit. 2013. *A Crisis of Culture: Valuing Ethics and Knowledge in Financial Services* (London: EIU).

76. Gilbert, M. 2014. Q&A with Richard Bistrong, the FCPA blogger who knows. *FCPA Blog*. **https://www.linkedin.com/pulse/article/20141122132021-310520166-q-a-with-richard-bistrong-the-fcpa-blogger-who-knows.** This interview originally appeared on *Corporate Compliance Insights*, **www.corporatecomplianceinsights.com** and is quoted with permission of CCI and Richard Bistrong, **www.richardbistrong.com**

77. Gilbert, M. (op. cit.).

78. Barnes, C. 2015. The ideal work schedule, as determined by circadian rhythms. *Harvard Business Review*, 01/2015.

79. Fox, T. 2012. How do you change a corporate culture. *FCPA Compliance and Ethics Blog* 2012. **https://tfoxlaw.wordpress.com/2012/11/08/how-do-you-change-a-corrupt-culture/**

80. Dietz, G. and Gillespie, N. 2012. Rebuilding trust: how Siemens atoned for its sins. *Guardian Sustainable Business*. **http://www.theguardian.com/sustainable-business/recovering-business-trust-siemens**

81. Siemens, 2015. **http://www.siemens.com/sustainability/en/core-topics/collective-action/integrity-initiative/index.php**

82. One Stone Advisors (op. cit.) and TeliaSonera material, including 2013 Interview Michaela Ahlberg, new Chief Ethics and Compliance Officer. **http://www.teliasonera.com/en/newsroom/news/2013/interview-michaela-ahlberg-chief-ethics-and-compliance-officer/**

83. One Stone Advisors. 2015. Interview with Dr Anna Young-Ferris, Leader UN Principles for Responsible Investment (UNPRI) Asia Pacific Academic Network and Lecturer University of Sydney Business School.

84. One Stone Advisors (op. cit.).

85. One Stone Advisors (op. cit.).

86. Hanson, K. 2014. *Business Ethics MOOC: Creating an Ethical Corporate Culture*. Markkula Center for Applied Ethics, Santa Clara University. **http://www.scu.edu/ethics/practicing/focusareas/business/moocs.html**

87. LaMotte, S. 2015. Employee engagement depends on what happens outside of the office. *Harvard Business Review*. **https://hbr.org/2015/01/employee-engagement-depends-on-what-happens-outside-of-the-office/**

88. One Stone Advisors (op. cit.).

89. The Ritz-Carlton Leadership Centre. 2014. *The upside of daily line-up*. 3 December. **http://ritzcarltonleadershipcenter.com/2014/12/upside-daily-line-up/**

90. Price, M. (op. cit.).

91. One Stone Advisors. 2015. Interview with Helena Hagberg, Special Advisor Sustainable Business, Swedish Ministry of Finance.

92. Carney, M. 2015. *Building real markets for the good of the people*. Bank of England speech, Lord Mayor's Banquet for Bankers and Merchants of the City of London at the Mansion House, London, 10 June. **http://www.bankofengland.co.uk/publications/Documents/speeches/2015/speech821.pdf**